TROPICAL ORCHIDS OF MALAYSIA & SINGAPORE

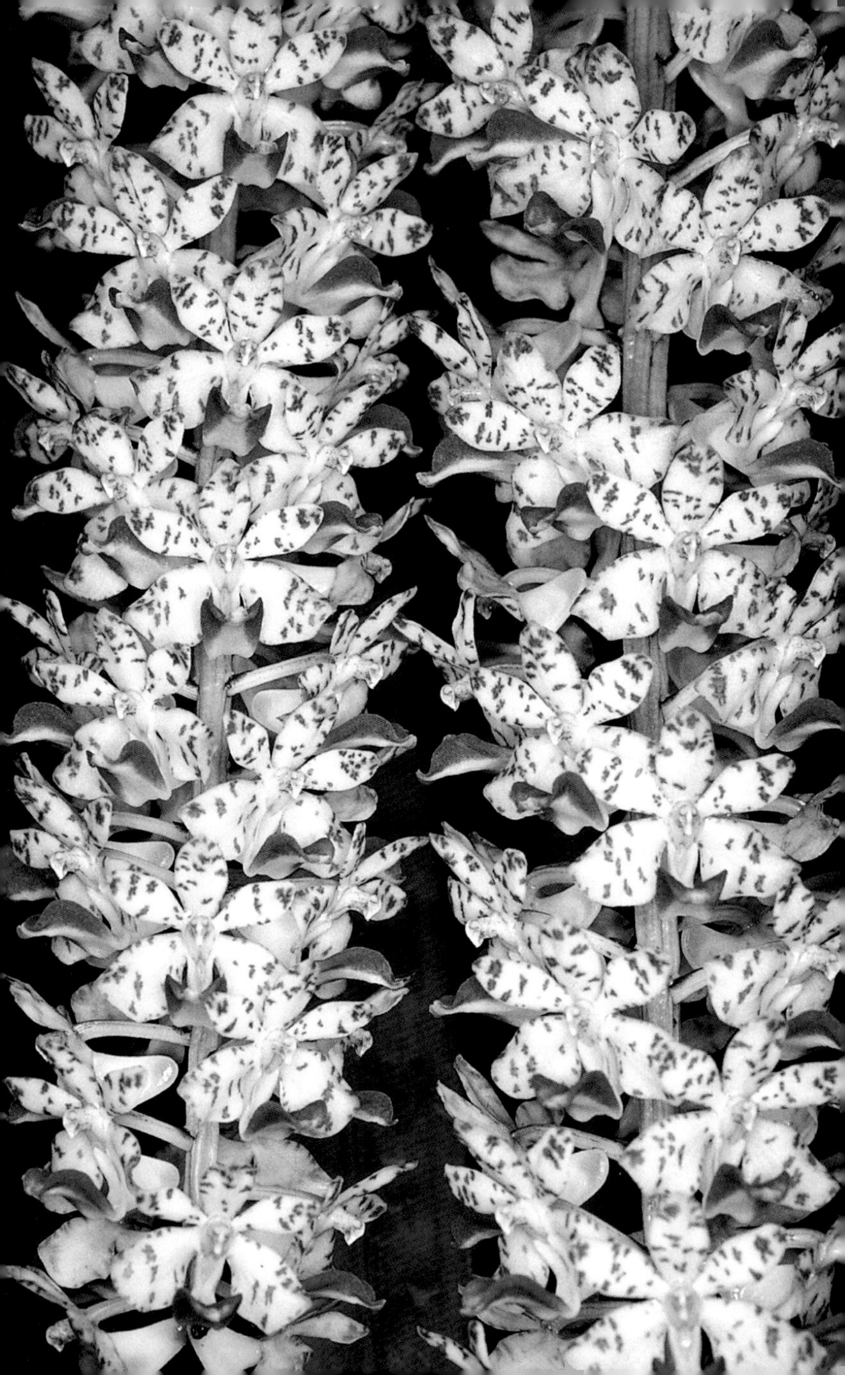

PERIPLUS NATURE GUIDES

TROPICAL ORCHIDS

of Malaysia & Singapore

Text and photographs by David P. Banks

PERIPLUS

EDITIONS

Published by Periplus Editions (HK) Ltd.

ISBN 962-593-156-2

Publisher: Eric M. Oey
Design: Peter Ivey
Editors: Rod Ritchie and Julia Walkden
Production: Rod Ritchie and Agnes Tan
Printed in Singapore

Distributors
Indonesia
PT Wira Mandala Pustaka
(Java Books - Indonesia),
Jalan Kelapa Gading Kirana,
Blok A14 No. 17,
Jakarta 14240

Singapore and Malaysia
Berkeley Books Pte. Ltd.,
5 Little Road #08-01, Singapore 536983

United States
Charles E. Tuttle Co., Inc.,
RRI Box 231-5, North Clarendon,
VT 05759-9700

About the Author

David P. Banks has been growing orchids for over 25 years. He is a Past President of the Orchid Species Society of New South Wales, and of the Parramatta and District Orchid Society, and has also served two terms as Vice President of the Orchid Society of New South Wales. He is a qualified orchid judge with the OSNSW as well as the Australian Orchid Council and is the current Editor of *The Orchadian* and the *Australian Orchid Review*. David's acclaimed photographs appear in many journals and he lectures extensively throughout Australia and overseas at workshops, society meetings and orchid conferences.

Acknowledgements

Most of the orchids illustrated in this book were grown by the author. However, I wish to thank the following people for allowing me to photograph plants in their collections; Margaret Barrett, Kerrie Bennett, Adrian Browne, Jim Cootes, Howard Gunn, Ray Hill, Sandy Holmes, Barry Long, Bruce Lonnon, Bob McCue, Brian Milligan, Jim Neal, Andy Phillips, John Roberts, Ken Smart and Darryl Smedley. I wish to express thanks to my photographic mentor, David Titmuss and to Jim Cootes who read and commented on the manuscript. Finally, and most importantly, I would like to thank my parents, Lynette and Graeme, for their encouragement.

This book is dedicated to the memory of my late wife Jannine Louise Banks.

Orchid on page 2: *Rhynchostylis retusa*

Introduction

Orchids have always held a fascination for people; there is an air of mystery surrounding them. Mention the word "orchid" and terms such as exotic, rare, expensive, beautiful, colourful and unique will arise. Everyone will have a different perception and all of these, no matter how diverse, will be correct.

Over 30,000 different orchid species are found on our planet, and well over 100,000 hybrid strains have been artificially propagated. Many of these hybrids are important commercial plants that are used as cut flowers and pot plants to satisfy strong demand, both locally and internationally.

What makes orchids different from other flowering plants? Their floral structure is obviously different. Orchids have three petals and three sepals which may, or may not, be alike. However, the third petal, known as the lip or labellum, is almost always highly modified and quite different from the other segments in size, shape and often colour. This structure often plays an important part in orchid pollination. Another feature, which helps to define orchids, is an appendage, known as a column, often located in the centre of the flower. The fleshy column combines the male (anther) and female (stigma) sexual parts on one structure. The anther and stigma are situated close together.

Orchid flowers use their form, colour and often fragrance to attract many creatures which act as pollinators. Bees, wasps, butterflies, moths, flies, beetles, ants and even birds assist the fertilisation of orchid flowers. Numerous orchid flowers actually mimic insects. Some deceive bees by looking like other flowers that offer a reward, while other species, often with deep maroon flowers, smell like rotten flesh, to attract flies!

Two main growth structures are found in orchids; monopodial and sympodial. Sympodial orchids, such as *Bulbophyllum* and *Dendrobium*, have a main stem, or pseudobulb, which is produced annually and matures at the end of each growing season, often culminating with flowering. During the next season, a new pseudobulb grows from the base of last season's growth. These pseudobulbs, which hold water and nutrients, are produced along a structure known as a rhizome. Monopodial orchids such as *Aerides*, *Phalaenopsis* and *Vanda* have main stems which grow constantly. These main stems produce flower spikes, correctly termed inflorescences, from or opposite the leaf axil.

Orchid seeds are minute and dust-like—a single fruit or capsule has the potential to produce up to a million seeds, depending on the species. However, orchid seeds have little food storage and rely on a specific type of fungus for their germination and development. The mortality rate in the wild is enormous. Now, orchid seeds are germinated in laboratories using a synthetic agar solution. This technique has made it possible to produce large quantities of both species and hybrids in a sterile environment. Depending on the genus, orchids can take from twelve months to twelve years from germination to

Amesiella monticola

Ascocentrum pumilum

flowering. On average, most orchids flower around four years from germination after starting out in life as green blobs known as protocorms.

One fact that quickly becomes apparent is that few orchids have "common" names. Many are simply referred to by their generic name. Many other groups of plants are known by their botanical names, as common names, often without the users knowing it. Such tongue-twisting examples include *Agapanthus, Bougainvillea, Chrysanthemum, Jacaranda* and *Rhododendron*. These names are derived from both Latin and Greek, however, Latin has been used extensively for scientific terms. This enables a universal system for communication. A problem with common names is that one name may refer to many completely different plants. Also, what may be a "local" name in one area may not be pertinent in another.

Over 75 per cent of the world's known orchid species occur in the tropics. This is a fact that doesn't surprise many. What may surprise, however, is that the majority of these "tropical" species (over 80 per cent) inhabit the cooler mountainous forests, at altitudes of over 1000 metres above sea level. These montane forests provide homes for an amazing number of diverse and unique plant and animal species. Isolated high peaks, such as Mt. Kinabalu in Sabah, have a high percentage of endemic species that are found nowhere else in the world.

In the tropics, most orchids are epiphytes, that is, they grow on trees for support and light. They are not parasitic, as they do not take food from the tree. Some species grow exclusively on rocks, and are known as lithophytes. Species that grow in the ground are referred to as terrestrials. Most of the tropical terrestrial orchid species are evergreen, unlike the deciduous terrestrials, which are generally found in more temperate climates. After flowering,

Dendrobium devonianum

Paphiopedilum druryi

these plants die down to storage organs, known as tuberoids, which have a similar life-cycle to bulbs.

A small percentage of orchids are saprophytes; these live off dead or decaying matter. There are even two Australian orchid species which grow and flower completely underground!

Only a small percentage of the world's orchid flora is cultivated in specialist nurseries, botanical institutions and private collections. Whilst the collection of popular species from the rainforests can threaten the survival of wild orchid populations, their main enemy is habitat destruction. Southeast Asia is one of the world's most densely populated regions, and its virgin forest continues to be cleared and burnt. New species continue to be discovered, and a number of "lost" species have been relocated. Yet one wonders how many species, both plant and animal, have become extinct before they have even been recorded.

Fortunately, most of the horticulturally attractive orchid species are entrenched in cultivation and have been propagated by division or by seed. The species at real risk are the countless miniatures, or "botanicals", as they are known. These have little commercial value and are of most interest to botanists and orchid species enthusiasts.

Hopefully, this book will dispel the myth that most orchid species look the same. In fact, these wonderful plants come in a huge array of shapes, sizes and colours, and are highly specialised. The tropical orchid species shown of the following pages represent the tip of an enormous iceberg.

Orchids will continue to captivate future generations with their unique beauty—indeed they are the monarchs of the plant kingdom.

—David P. Banks

Aerides

Opposite

Top left:
Aerides odorata var. *alba*

Top right:
Aerides quinquevulnera

Bottom left:
Aerides rosea

Bottom right:
Aerides quinquevulnera var. *purpurata*

The name *Aerides* literally means air-plant. About 20 species belong to this genus of monopodial epiphytes. Many of these sturdy plants are found in the warm lowlands of the tropics. Most species are easy to grow and have highly perfumed flowers, which has made them popular in horticulture. ***Aerides odorata*** is variable in colour, from deep pink to pure white, and is widely distributed throughout Southeast Asia. Albinistic or white-flowered forms of these species are highly prized by collectors. ***Aerides quinquevulnera*** is from the Philippines and New Guinea and has waxy pink, spotted to deep purple flowers which can last for over a month in pristine condition. ***Aerides rosea***, from northern Thailand, was formerly well known as *Aerides fieldingii*. It has long, pendent inflorescences of lolly-pink flowers, with some darker spotting on the petals. *Aerides* species are frequently cultivated, attached to the trunks of trees which will not shed their bark. In optimum conditions, the thick white roots will attach firmly to the host and ramble quite a distance from the plant.

Many hybrids have been made between *Aerides* and other members of what are loosely termed the "vandaceous" family. Such combinations include *Aeridocentrum* (x *Ascocentrum*), *Aeridopsis* (x *Phalaenopsis*), *Rhynchorides* (x *Rhynchostylis*) and *Aeridovanda* (x *Vanda*). The genus *Christieara* is a three-way hybrid involving *Aerides*, *Ascocentrum* and *Vanda*. Many of these hybrids come in a wide range of colours due to the high degree of genetic diversity. They also have long-lived flowers, a trait which has made them popular as cut flowers and as flowering pot plants. Some of the other *Aerides* species in cultivation include *A. lawrenciae* (which is like a giant *A. odorata*), *A. crassifolia* (a fleshy plant with large purple flowers), *A. krabiense* (deep mauve flowers), the lilac and cream *A. falcata*, and its close relative *A. houlletiana*, which has yellow, white and purple blooms.

Amesiella and Ascocentrum

Opposite

Top left:
Ascocentrum ampullaceum

Top right:
Ascocentrum garayi

Centre left:
Ascocentrum ampullaceum var. *album*

Bottom:
Amesiella philippinensis

Amesiella

The Philippine species ***Amesiella philippinensis*** was known for many years as *Angraecum philippinensis*. The genus *Angraecum* is actually restricted to Africa and Madagascar. There are two species in this monopodial genus, with a second species recognised in 1998. Both species are epiphytes which have short spikes bearing up to five disproportionately large crystalline white flowers. ***Amesiella philippinensis*** has a prominent yellow blotch on the labellum and occurs at lower altitudes. The more robust ***Amesiella monticola*** has larger pure white flowers complete with a differently-shaped, three-lobed labellum and longer spur. This spur is filled with nectar to attract its pollinator, which is probably a moth since the flowers are fragrant at night.

Ascocentrum

The monopodial genus *Ascocentrum* is a group of small, compact plants consisting of less than ten epiphytic species which have played an important part in the development of vandaceous hybrids. *Ascocentrum* has been bred with members of the genus *Vanda* to create the artificial genus *Ascocenda*. The *Ascocentrum* has helped introduce bright colours to its progeny, while also reducing the plant size.

Ascocentrum ampullaceum has round, glowing purple flowers and often flowers profusely. Recently, a very rare pure white clone of the species was discovered. White-flowered forms of coloured orchids always demand high prices due to their scarcity. Numbers can be increased by division of older plants, seed and tissue culture. Plants produced by tissue culture, called mericlones, are genetically identical to the parent plant, whereas seedlings may display variation in colour and form. The bright orange ***Ascocentrum garayi*** is a popular plant in its native Thailand. For years this species has been confused with the related *Ascocentrum miniatum*. The minature ***Ascocentrum pumilum*** is endemic to Taiwan.

Bulbophyllum

Below:
Bulbophyllum longiflorum

Opposite

Top left:
Bulbophyllum blepharistes

Top right:
Bulbophyllum alsiosum

Bottom left:
Bulbophyllum careyanum

Bottom right:
Bulbophyllum dayanum

Over 2000 named species are included in this cosmopolitan genus, with flowers that come in all shapes, sizes and colours. Species are still being discovered, and many others have not been formally named by botanists. Most of these sympodial plants grow as epiphytes and lithophytes. A majority of species produce a cylindrical pseudobulb with a single leaf, which develops along a creeping rhizome. This feature helps identify the genus out of flower. *Bulbophyllum* includes some of the world's smallest orchids, plus other species which grow to a massive size. The flowers are generally very "un-orchid" like, and are highly specialised to attract their specific pollinators.

Bulbophyllum alsiosum occurs in the lowland forests of the Philippines, and has single golden-brown flowers. This orchid is related to the widespread ***Bulbophyllum lobbii***. This robust species is common throughout humid forests in Southeast Asia, with some colourful forms found in Borneo and Indonesia.

Below:
Bulbophyllum weberi

Opposite

Top left:
Bulbophyllum facetum

Top right:
Bulbophyllum lasiochilum

Bottom left:
Bulbophyllum lobbii

Bottom right:
Bulbophyllum macranthum

Bulbophyllum facetum is a recently described species restricted to a small area of montane forest in central Luzon, in the Philippines. The mottled flowers of this impressive species open during the day and close late in the afternoon. This ritual is repeated for only a few days.

Bulbophyllum blepharistes is a two-leaved species from Thailand which is unique. The plant habit resembles some of the *Bulbophyllum* species which occur in Africa and Madagascar.

Bulbophyllum careyanum is found from India through to Thailand. Individual plants may produce flowers which range in colour from an attractive yellow-orange to a dull maroon-brown. These are produced in tight clusters of up to 30 smelly flowers.

Bulbophyllum dayanum, from Myanmar and Thailand, has flowers that look like little fuzzy monsters! They are produced in short spikes at the base of the plant, which often has a purple tinge.

Bulbophyllum lasiochilum has single crab-like flowers which range from in colour from orange-brown to deep maroon. It is found throughout Southeast Asia, as is ***Bulbophyllum macranthum***.

One of the more showier species is ***Bulbophyllum pectinatum***, found in cooler mountainous forests from

Bulbophyllum 15

Opposite

Top left:
Bulbophyllum pectinatum

Top right:
Bulbophyllum wendlandianum

Centre left:
Bulbophyllum medusae

Bottom:
Bulbophyllum medusae

Myanmar, to Vietnam and southern China. A pure green form is endemic to Taiwan.

Bulbophyllum medusae takes its name from the woman in Greek mythology who had snakes as hair! This lowland plant, found throughout tropical Southeast Asia, has clusters of flowers varying in colour from pure white to creamy tones, with light-brown pepper spotting.

Another species from Thailand is ***Bulbophyllum wendlandianum***. This species is a member of the *Cirrhopetalum* section of *Bulbophyllum*. This section is generally characterised by having flowers produced in an umbel, lateral (lower) sepals which are fused, and filaments and appendages on the dorsal (upper) sepal and petals. These "flags" move in the slightest breeze and help attract potential pollinators. Some species of *Bulbophyllum* only flower in response to wet and dry seasons, while others flower throughout the year.

Bulbophyllum longiflorum has one of the most widespread distributions of the tropical epiphytic orchids. It is another member of the *Cirrhopetalum* section and has been found growing at altitudes from 200 to 1400 metres in central and southern Africa, Madagascar, the Seychelles, the Reunion Islands, Mauritius, Borneo, the Philippines, New Guinea, Guam, northeastern Australia, the Solomon Islands, New Caledonia, Fiji, Samoa and Tahiti. It is quite a traveller! The colour varies quite a bit, with the most common colour being depicted in the pictured clone from the Philippines. In suitable conditions, this species can grow into very large plants which flower intermittently throughout the year.

Bulbophyllum weberi is a closely related species which is endemic to the Philippines.

Chiloschista

Below:
Chiloschista phyllorhiza

Opposite

Top left:
Chiloschista parishii

Top right:
Chiloschista ramifera

Bottom left:
Chiloschista phyllorhiza

Bottom right:
Chiloschista trudelii

This bizarre group of plants is distinctive in that it has no leaves. The mass of flattened roots, which ramble over its host, have a grey-green tinge and actually perform the vital process of photosynthesis. Pendent flower spikes appear from the growing point. The flowers are quite showy and often amaze those seeing them for the first time. About 25 species have been documented, but some of these may prove to be mere colour variations.

The spotted ***Chiloschista parishii***, the boldly blotched ***Chiloschista ramifera***, and the mustard-coloured ***Chiloschista trudelii*** are all from Thailand. They were formally described in the late 1980s. However, it is impossible to differentiate between them out of flower. All of these species have pendent sprays of waxy, circular flowers. A large plant in full bloom is an impressive sight. ***Chiloschista phyllorhiza***, from northern Australia, differs in floral structure and in having blooms which only last a day. As a result, this plant may be transferred at some stage to a new genus.

Above:
Coelogyne chloroptera

Opposite

Top left:
Coelogyne pandurata

Top right:
Coelogyne calcicola

Centre left:
Coelogyne radicosa

Centre right:
Coelogyne lentiginosa

Bottom:
Coelogyne mooreana

Coelogyne

Coelogyne is a large, diverse genus of sympodial orchids with close to 200 described species, almost half of which are in cultivation. Of Asian origin, these species are found from Nepal to China and Malaysia, the Philippines, Indonesia, Papua New Guinea and the Pacific islands. The plants produce one or two leaves, and often flower from pendent inflorescences produced with the new growths. Most of the members of this showy genus of epiphytes and lithophytes have white or green flowers, with contrasting labellums displaying many brown markings in them.

Some species, such as ***Coelogyne pandurata***, inhabit the lowlands of Malaysia and Borneo, often forming massive clumps on tree trunks along river banks next to rainforest. This species is popular in cultivation because of its large, green flowers which have an almost black, violin-shaped labellum. ***Coelogyne lentiginosa***, from Myanmar, Thailand and Vietnam, has short sprays of creamy flowers. The flower has a contrasting, flared, brown and white labellum.

The Philippine endemic, ***Coelogyne chloroptera*** has up to a dozen green flowers on upright inflorescences. The markings on the labellum are brown and cream. ***Coelogyne radicosa*** is a very rare Malaysian species. It displays its distinctive orange to brown flowers, with attractive broad labellums, at intervals throughout the year. One of its unique features, rare in the genus *Coelogyne*, is its ability to continue flowering off the same inflorescence for a number of seasons. After producing a few flowers, the thin inflorescence appears to go dormant, but stays green. Some months later it will elongate and again produce blooms.

Many *Coelogyne* species are also found in the cooler mountainous regions; most of these have white flowers. ***Coelogyne calcicola***, still scarce in cultivation, is a robust species with a long rhizome between the two-leaved pseudobulbs. The queen of this genus is arguably ***Coelogyne mooreana***, a rare, cool-growing epiphytic species from the mountains of Vietnam.

Above:
Cymbidium dayanum

Opposite

Top left:
Cymbidium bicolor

Top right:
Cymbidium insigne var. *album*

Bottom:
Cymbidium lowianum

Cymbidium

The genus *Cymbidium* has some fifty or so species distributed throughout Asia and south to Australia. These plants have been cultivated in China and Japan for centuries, where they are also valued for their spiritual and medicinal purposes. Most of the species are terrestrial, with upright flower spikes bearing blooms in many colours. In the lowlands, most cymbidiums take to the trees as epiphytes, growing in high light. Many of these species have long pendent inflorescences and thick, leathery leaves.

Countless hybrids have been artificially created over the past century. Today these hybrids form the basis of an important cut flower industry in temperate climates. Some of the more compact plants are also sold as flowering pot plants. These are popular, as they provide a floral arrangement for up to eight weeks—a lot longer than a bunch of flowers. The added bonus is you get the opportunity to try to it bloom it again!

The terrestrial ***Cymbidium insigne*** has been one of the most influential species in *Cymbidium* hybridising. It is native to Thailand, Vietnam and China, and has light pink to white flowers on erect spikes. ***Cymbidium lowianum*** is another hardy terrestrial species, with very long arching spikes of up to 30 olive green flowers, with a contrasting cream and red lip. It is distributed from Myanmar, through Thailand to southern China. ***Cymbidium bicolor*** has thick, succulent leaves, and spends its life as an epiphyte, perched high on suitable trees in full sunlight. This variable species has a huge range, from Sri Lanka, India across to China, Malaysia, Indonesia, and the Philippines. ***Cymbidium dayanum*** has a similar lifestyle and distribution, but extends also to Taiwan and Japan. The variegated-leafed form is highly prized by collectors. Variegation is very rare in orchids, and because of this, command high prices for stable forms. These are most popular in Japan and to a lesser extent China, where these cultivars are grown purely for their foliage.

Above (top):
Dendrobium devonianum

Above (bottom):
Dendrobium capra

Opposite

Top left:
Dendrobium bullenianum

Top right:
Dendrobium chrysanthum

Bottom:
Dendrobium crumenatum

Dendrobium

The genus *Dendrobium*, with in excess of 1500 species, has always been popular with orchid growers. It enjoys a wide distribution, from India and Sri Lanka, through Southeast Asia to New Guinea, Australia and the Pacific Islands. This genus is so complex, that it is highly likely that it will be dissected by botanists into a number of smaller genera. This has already started among some of the Australasian species.

An amazing diversity of plant habit, flower form and colour is found in this large genus. Almost all colours and combinations are represented in its flowers. *Dendrobium* contains species whose individual blooms last for only a few hours, and others which can persist for up to nine months in pristine condition. Quite a number of species produce new plants off the older pseudobulbs. These are called aerials or "keikis"—a Hawaiian word which means baby. Once these growths have hardened off and produced roots, they can be removed and grown as a new plant.

Many hybrids have been developed in tropical countries, for both orchid enthusiasts and the cut flower industry. The majority of flowers marketed as "Singapore Orchids" are actually *Dendrobium* hybrids. These hybrids were bred to produce plants which continue to flower off the old pseudobulbs, as well as the new growths, throughout the year. Generally the flowers are shapely, and are from white to deep mauve. However other strains exist, with narrower segments, in colours from green, through to yellow, to rusty tones. They last well as cut flowers and are frequently seen as table decorations in hotels and restaurants. The white varieties are used in wedding bouquets.

Dendrobium inflatum, whilst common in Java, is scarce in cultivation. It produces short-lived, crystalline-white flowers in pairs. The broad labellum is also white, with a small greenish-yellow blotch.

Opposite

Top:
Dendrobium gonzalesii

Centre left:
Dendrobium cariniferum

Centre right:
Dendrobium dearei

Bottom left:
Dendrobium nobile var. *virginale*

Bottom right:
Dendrobium draconis

A widespread species, ***Dendrobium spurium*** has flowers lasting only one or two days.

Dendrobium crumenatum is known as the "Pigeon or Dove Orchid". This orchid is a common species in the tropics, and is distributed throughout Southeast Asia. It grows as an epiphyte, often in full sun. A sudden drop in temperature of about 10°C will induce flowering. Such an event takes place during tropical storms. Exactly nine days later, the plants will bloom profusely. However, these flowers only last one day. This event will be repeated many times during the season.

Dendrobium junceum, from the Philippines, has long pendent terete stems which are swollen at the base. It, too, has short-lived flowers which are produced singularly.

Dendrobium secundum, sometimes called the "Toothbrush Orchid", occurs throughout Southeast Asia. Numerous pink to purple flowers are produced on only one side of the inflorescence, and always face upwards. White forms are rarely encountered. These would be more common in cultivation than in the wild, as their numbers have been increased by tissue culture.

Dendrobium nobile is a variable species throughout its range from India to southern China. It varies in colour from deep purple to pure white, with many shades and bicoloured combinations in between. Many varieties of this easily grown orchid are under cultivation. This and related species have been used to create the thousands of "Soft Cane" *Dendrobium* hybrids. Primarily bred in Japan and Hawaii, these hybrids have subsequently been marketed throughout the world. In the dry season, the plants shed their leaves and are dormant. Once the rains come, the plants burst into flower and produce the next season's growth. This deciduous feature is common with many of the *Dendrobium* species which have evolved to adapt to distinct wet and dry seasons. Thereby the plants grow quickly while there is moisture available and bloom at the change of seasons. The successfully pollinated flowers produce seed capsules which wait until the following rains to ripen and release their dust-like seeds.

Dendrobium devonianum follows the same pattern, and has a similar distribution to the previous species. The plants actually look dead while dormant, because of their

Opposite

Top left:
Dendrobium jacobsonii

Top right:
Dendrobium lindleyi

Centre left:
Dendrobium stricklandianum

Bottom:
Dendrobium harveyanum

thin wiry pseudobulbs. They flower off these leafless canes before starting next season's growth. It is a delightful, tri-coloured species, being lavender and white, with yellowish-green markings on the fringed labellum.

Dendrobium chrysanthum flowers while in full leaf, then sheds the leaves immediately after flowering. This orchid has round, orange-enamelled flowers, with a couple of dark, blood-red blotches on the lip. This pendulous species can grow over a metre long, and it flowers along the nodes opposite the thin leaves. It occurs from India, across to Vietnam and southern China, where the dried stems are used in medicine.

The Philippines have a large array of species, and many of these are endemics. ***Dendrobium gonzalesii*** is a pendent grower from the mountains, whose flowers vary in colour from bluish-mauve, through pink to white. The related ***Dendrobium serratilabium***, as the name suggests, has a serrated labellum, as can be seen in the photo. It varies in colour from a dark, striped brown through shades of yellow to green.

Dendrobium bullenianum is from the lowlands and was formerly known as *Dendrobium topaziacum*. It has tall canes which often droop if unsupported. The small orange and red-striped flowers are produced in dense clusters along the leafless pseudobulbs.

Dendrobium dearei, despite its papery appearance, has long-lived, white flowers, with a green throat in the labellum.

Dendrobium uniflorum is a member of a group of species known as the "Popcorn Orchids". This species has a green and white flower that goes yellowish with age.

Dendrobium stricklandianum is from Taiwan, Japan and China. It was formerly known to orchid growers as *Dendrobium tosaense*. This floriferous species has light green blooms with a contrasting white, red and yellow labellum.

Both ***Dendrobium cariniferum*** and the closely related ***Dendrobium draconis*** are found throughout most of Southeast Asia. An interesting feature of these glossy, white-flowered species are the pseudobulbs which are densely covered in fine black hairs.

Opposite

Top left: *Dendrobium secundum*

Top right: *Dendrobium secundum* (white form)

Centre left: *Dendrobium sanguinolentum*

Bottom: *Dendrobium nobile*

Dendrobium lindleyi is a popular species in cultivation, often grown under its synonym, *Dendrobium aggregatum*. Common in Thailand, this epiphytic species is found from India to China. The pendent sprays of lemon-yellow to orange flowers are quite large, considering its compact growth habit. It has a small, furrowed, one-leafed pseudobulb, and the flower spikes come from nodes halfway along this storage organ.

Dendrobium harveyanum is readily identified in flower because of its unique petals which have long filaments on the margins. The round labellum is densely fringed. Up to eight canary-yellow flowers develop from short sprays near the top of the matured growths. This is a rare species found in Myanmar, Thailand and Vietnam.

Dendrobium pulchellum, previously known for many years as *Dendrobium dalhousianum*, is found in Nepal, northern India, Myanmar, Thailand and southern China. It can produce very long, cane-like pseudobulbs, up to two metres tall. Up to a dozen cream to apricot large flowers, with a couple of dark maroon blotches on the lip, are produced on pendent spikes off the older leafless stems. These can produce blooms for many years.

Dendrobium sulcatum has a similar natural range. It has flattened pseudobulbs and golden flowers with a network of red veining in the labellum.

Dendrobium sanguinolentum, despite its distribution, is rarely seen. It has been recorded in Thailand, Malaysia, and Indonesia. This epiphytic species can grow over a metre tall, often in exposed locations at moderate elevations. The round flowers are white to cream, with attractive splashes of purple on the tips of all the floral segments.

Dendrobium jacobsonii is a high mountain plant found only in eastern Java. It has bright, almost glowing, scarlet-red flowers. A short-growing species, the pseudobulbs are seldom over 30 cm long. Many high altitude *Dendrobium* species from Indonesia and New Guinea have long-lived, brightly-coloured flowers. Unfortunately, when the plants are brought down to the lowlands, they cannot cope with the constant heat and humidity, and soon perish.

Below (bottom):
D. serratilabilum

Opposite

Top left:
Dendrobium inflatum

Top right:
Dendrobium tobaense

Centre left:
Dendrobium spurium

Bottom:
Dendrobium sulcatum

Another species endemic to east Java, but this time from the lowlands, is ***Dendrobium capra***, with spikes of up to 16 shiny, light olive-green flowers, which open widely.

Dendrobium tobaense, from northern Sumatra, is a new species to science, having been described for the first time only in 1993. This species has a bizarre flower, unlike any other *Dendrobium*, with segments which are green with darker veins, and a narrow labellum which is predominantly red and white. Plants of this species have only recently entered cultivation, and will be propagated from seed to increase their numbers.

Dendrobium uniflorum

Dendrobium junceum

Dendrobium pulchellum

Dendrochilum

Below:
Dendrochilum tenellum

Opposite

Top left:
Dendrochilum arachnites

Top right:
Dendrochilum cobbianum

Bottom left:
Dendrochilum tenellum

Bottom right
Dendrochilum cootesii

Members of this genus of over 250 species have been affectionately known by the indigenous people as "Rice Orchids". They are sympodial, have a single leaf, and are related to *Coelogyne*. The majority are epiphytes, which grow in the mossy cloud forests in mountainous regions, where significant temperature extremes rarely occur. Only a few species make their homes in the tropical lowlands. They bloom once a year, with the developing new growth, and the flowers are arranged in two distinct rows. The centre of distribution for the genus *Dendrochilum* is the Philippines, with numerous species found in Borneo and Sumatra. Until recently, little literature was available on these miniature orchids, which resulted in many plants masquerading under incorrect names. ***Dendrochilum tenellum*** looks more like a grass than an orchid. It is a common species found throughout the Philippine archipelago, often forming large clumps on moss-covered rainforest trees.

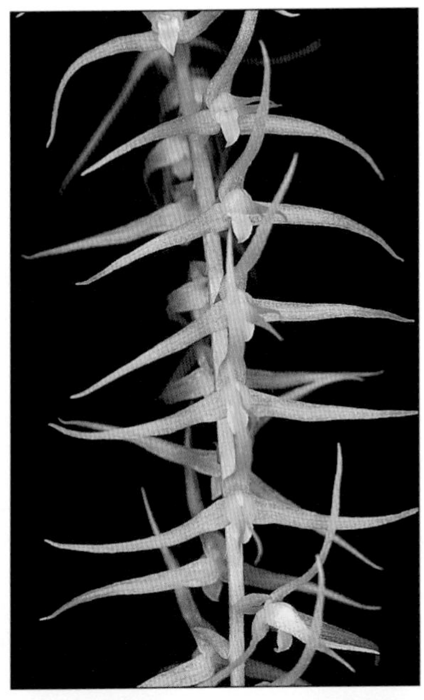

Dendrochilum 35

Below
Dendrochilum wenzelii

Opposite

Top left:
Dendrochilum javierense

Top right:
Dendrochilum saccolabium

Bottom left:
Dendrochilum wenzelii

Bottom right:
Dendrochilum uncatum

Dendrochilum uncatum has green to yellow blooms.

The following six species are all endemic to the Philippines. ***Dendrochilum cobbianum***, the most frequently encountered species, varies in colour from white through to yellow and green, and sometimes the labellum is a contrasting colour. One distinctive feature of this species is the presence of a nectar gland in the centre of the labellum. Red flowers are rare in this genus.

While the colours of ***Dendrochilum saccolabium*** and ***Dendrochilum wenzelii*** are similar, their flowers are shaped differently. They also have fine grass- or rush-like foliage.

Dendrochilum javierense is similar to the latter, and has yellowish green flowers, resembling birds in flight. The true ***Dendrochilum arachnites*** is quite scarce, although easily recognised by its spidery, greenish-yellow flowers and rambling growth habit, with pseudobulbs spaced along the rhizome.

The rare ***Dendrochilum cootesii*** was only formally described in 1997, and is a distinctive new addition to the Philippine orchid flora.

Above:
Eria coronaria

Opposite

Top left:
Eria javanica

Top right:
Eria longissima

Bottom:
Eria vanoverberghii

Eria

The genus *Eria* is widespread throughout tropical Asia, to New Guinea, Australia and Polynesia. Numerous delightful examples are found within a catalogue of over 500 species, however very few are in cultivation. Within this sympodial genus, which is related to *Dendrobium*, are an enormous number of distinct plant forms. Most species have small flowers, but they often compensate for this by producing prolific numbers. The individual flowers last for less than a week, however, plants often bloom a number of times a year. They generally occur as epiphytes, although some are lithophytes.

Eria coronaria is arguably the most regal member of the genus. This orchid is found from the Himalayas to Vietnam and southern China. Its flowers are very similar to some of the *Dendrobium* species from Australia. ***Eria javanica*** has starry, cream flowers on long spikes, and has a wide distribution from India, through Southeast Asia, to New Guinea. The following two species, although quite different, are endemic to the Philippines. ***Eria longissima*** has cane-like pseudobulbs, which can grow to over two metres in height. Flowering is initiated after a sudden drop in temperature, during a thunderstorm. The plants will quickly develop buds, before bursting into flower about a week later, but they only last for a couple of days. This ritual is repeated many times during the season. The flowers are fragrant. ***Eria vanoverberghii*** produces a number of spikes from the top of the matured pseudobulb, and flowers sequentially over a couple of months. This species has conspicuous, white bracts behind the flowers.

Most *Eria* species are easy to cultivate. In the tropics, many lowland species can be attached to suitable trees that don't shed their bark. They like to grow in dappled light. Some of the smaller growing species perform best in pots, grown in a bark-based medium, and kept moist whilst the plants are in active growth.

Eria 39

Liparis

Opposite

Top left:
Liparis gibbosa

Top right:
Liparis plantaginea

Bottom:
Liparis latifolia

The genus *Liparis* is cosmopolitan, with about 250 species, a high percentage of these being terrestrial. In the tropics, however, most members are epiphytes. Often the flowers are in various shades of yellowish green, although some species have contrasting bright orange and red labellums. These flowers have developed a reputation amongst orchid growers for having quite a putrid fragrance, like a dog that hasn't been bathed for months! This is obviously a modification to attract a pollinator.

Liparis latifolia, frequently misidentified as *Liparis nutans*, is widespread throughout Southeast Asia and has quickly become entrenched in cultivation. This plant has a distinctive coppery appearance, with up to 40 dark brownish-red flowers arranged on an arching inflorescence. ***Liparis plantaginea*** grows as an epiphyte or sometimes a terrestrial, and occurs from India to Vietnam. The translucent, bottle-green flowers are distinctive, and appear with the new growth. ***Liparis gibbosa*** is a miniature epiphyte found in Myanmar, Thailand, Malaysia and the Indonesian island of Java. It flowers sequentially, with one or two out at any given time. Flower colour varies from yellowish green to orange and reddish brown. *Liparis* species are often only seen in specialist collections or botanical gardens.

Most *Liparis* species have soft leaves which will burn if exposed to full sun. They are mostly found in shady environments, often near creeks, where there is always high humidity. The epiphytic species often grow on the moss-covered limbs or trunks of trees on the edges of rainforest. *Liparis* species are quick-growing subjects in cultivation if their native environment is simulated. It needs to be kept moist, shaded and with circulating fresh air. Most do well in small pots of a bark-based medium, which drains freely.

Above (top):
Paphiopedilum callosum

Above (bottom):
Paphiopedilum hookerae (top) & P. volonteanum

Opposite

Top left:
Paphiopedilum bellatulum

Top right:
Paphiopedilum concolor

Centre left:
Paphiopedilum acmodontum

Bottom:
Paphiopedilum charlesworthii

Paphiopedilum

The "Slipper Orchids" have long been highly prized in horticulture. They are cultivated throughout the world, and countless hybrids have been derived from the 80 or so different species. The range of the genus *Paphiopedilum* extends from India, eastwards across southern China to the Philippines, throughout Southeast Asia and Malaysia, to New Guinea and the Solomon Islands. New species continue to be discovered, particularly in remote rainforest areas of Kalimantan, Indonesia and China.

A huge amount of diversity is found within the slipper orchids. Some species are terrestrial, growing through the leaf litter on the forest floor, others are lithophytes which show a preference for limestone cliffs, while a number are epiphytes, and live in the major forks of suitable trees. Most of the species produce a single flower. Some may have up to a dozen or more open at one time, while others flower sequentially. These flowers, which come in a wide range of colour and form, often last for well over a month in pristine condition.

Most of the species have plain, green strap leaves, yet others have a distinctive, mottled foliage which makes them attractive even when out of flower. They generally grow in quite moist and humid environments. These orchids do not have pseudobulbs, but store water in their fleshy leaves and thick, hairy root systems.

Unfortunately, slipper orchids have been over-collected in the wild and many species are on the brink of extinction in their native habitat. Forest fires can also destroy remote populations. Often the plants occur in vast numbers in a relatively small area, which makes the plants vulnerable. Numerous documented cases show that entire colonies of plants have been completely decimated by collectors in a single day. Some species may in fact be extinct in the wild. For example, ***Paphiopedilum druryi***, known only from the Travancore Hills in southern India, was collected to near extinction.

Opposite

Top left: *Paphiopedilum dayanum*

Top right: *Paphiopedilum druryi*

Bottom: *Paphiopedilum glaucophyllum*

The majestic ***Paphiopedilum rothschildianum***, from Mt. Kinabalu in Sabah, East Malaysia, may have suffered the same fate. Up to five dark, striped flowers are produced on an upright spike, and each flower is up to 30 cm across the extended petals. It has only ever been known from a few sites within the Mt. Kinabalu National Park. Despite this, orchid poachers have infiltrated its home and collected all of the plants.

Fortunately, the Asiatic slipper orchids are fairly easy to propagate from seed, and are firmly entrenched in cultivation by specialist nurseries and private collections. Most of the horticulturally attractive species have been multiplied to a point that the price has dropped, which has satisfied the growing world-wide demand for these unique plants. Another species, which has only been recorded from the lower slopes of Mt. Kinabalu, is ***Paphiopedilum dayanum***. This green-striped and maroon, single-flowered species has the added attraction of tessellated foliage.

Paphiopedilum hookerae has always been a rarity. It is endemic to Sarawak and western Kalimantan in Indonesia. For many years it was considered to be one of the rarest of the slipper orchids. The even rarer and closely related ***Paphiopedilum volonteanum*** has only been found in Sabah, East Malaysia.

For many years, ***Paphiopedilum delenatii***, from Vietnam, was only known from a handful of collected plants which were sent to France before the First World War. All but one of these plants died. The survivor was nurtured by a French nursery, and subsequently mass-produced over many years from seed. It has a pale pink flower, with a darker labellum or "slipper", produced singularly or in pairs. This orchid has been propagated to the point that around the world there now would be tens of thousands of orchids, all derived from that single plant—truly, one of horticulture's major success stories. Ironically, this species was rediscovered in Vietnam in the early 1990s.

Nine species of slipper orchids are found in the Philippines, and all but one are endemic. The multi-coloured ***Paphiopedilum acmodontum*** was discovered in the late 1960s and has become popular in cultivation. It is closely related to ***Paphiopedilum callosum***, which is found in Thailand, Cambodia, Laos and Vietnam. This is an impor-

Opposite

Top left: *Paphiopedilum hirsutissimum* (normal and albino forms)

Top right: *Paphiopedilum rothschildianum*

Bottom: *Paphiopedilum primulinum*

tant species in the development of what are commonly known as "Maudiae" type *Paphiopedilum* hybrids. As well as the usual green, white and purple combination, there is a rare albino form, having white flowers with green stripes. These species grow as terrestrials in shaded situations among the leaf-litter on the fringes of rainforest.

Northeastern Thailand is the home of ***Paphiopedilum sukhakulii***, a stately species discovered in 1964. This species was collected in vast numbers after its discovery, putting immense pressure on the natural populations. There are also pure green forms of this species.

Some species of slipper orchid grow almost exclusively on limestone monoliths. Their roots cling tenaciously to the rock and often travel vast distances searching for moisture and nutrients along any cracks. Most of these species have white to cream round flowers, with various degrees of spotting on the flowers. The leaves are also mottled and give the impression of having a clear plastic coating.

Paphiopedilum bellatulum is only known from the northern border of Thailand and Myanmar. It has large, white flowers with large, deep maroon spots. The very short flower stem lets the single flower rest on the leaves. A prized albino form also exists, pure white without any spotting. This species is in the background of many of the complex *Paphiopedilum* hybrids. It imparts spots and round shape to its progeny, but, unfortunately, also a short stem.

A very similar species is ***Paphiopedilum concolor***. It is more widespread, found from Myanmar across to Vietnam and southern China. The cream to yellow flowers with fine purple dotting are often produced in pairs.

Paphiopedilum charlesworthii is one of the most distinctive species in the genus. It has only ever been found in Myanmar. Its outstanding feature is the large musk pink dorsal sepal. The petals and labellum are a glossy brown.

Paphiopedilum hirsutissimum was discovered last century and is distributed from the northeastern part of India to northern Myanmar and Thailand. It has very hairy flower stems and the green, brown and purple flowers are also densely covered in hairs. There is also a very rare pure green form of this species.

Below (left):
Paphiopedilum delenatii

Below (right):
Paphiopedilum sukhakulii

Opposite

Top left:
Paphiopedilum stonei

Top right:
Paphiopedilum supardii

Bottom:
Paphiopedilum victoria-regina

The multifloral slipper orchids are almost exclusively epiphytes or lithophytes and have stiff, plain green leaves. These are generally found in quite sunny situations.

Paphiopedilum stonei only occurs on limestone cliffs in western Sarawak, East Malaysia. Two to four flowers are produced off growths which take two years to mature. Two prominent features are the white dorsal sepal and the deep pink labellum which is always pushed forward. It is unlikely to be confused with any other species.

Paphiopedilum supardii is endemic to southeast Kalimantan in Indonesia, growing in leaf-mould over limestone boulders. Healthy, mature plants generally carry up to six flowers, although up to a dozen has been reported. The unusual flowers are complete with spots, dots, stripes, warts and hairs.

Paphiopedilum glaucophyllum is a lowland species from east Java, where it often grows on steep rock faces in low light.

Paphiopedilum victoria-regina from Sumatra is synonymous with *Paphiopedilum chamberlainianum*. Plants can grow quite large, with leaf spans of over one metre on plants in optimum conditions.

Paphiopedilum primulinum was discovered in 1972 and is restricted to northern Sumatra. It has bright clear yellow flowers, with a green tinge on the dorsal sepal. Its discovery caused a sensation when the original collector was led to its haunt by a vivid dream, which suggested the location of a blue slipper orchid.

Phalaenopsis

Below:
Phalaenopsis equestris

Opposite

Top left:
Phalaenopsis amabilis

Top right:
Phalaenopsis fimbriata

Centre left:
Phalaenopsis lindenii

Bottom:
Phalaenopsis javanica

Most people would be familiar with *Phalaenopsis,* known as the "Moth Orchids". Many hybrids have been produced from the 50 or so wild species. It is arguably the most important commercial genus of orchids in the world. Tens of thousands of flowering plants are sold annually throughout the world to cater for the growing "flowering pot plant" trade. White *Phalaenopsis* hybrids are still one of the most popular flowers throughout the world, and are often used in wedding bouquets. Most of these white hybrids have been derived from the magnificent "White Moth Orchid", ***Phalaenopsis amabilis***. However, *Phalaenopsis* do not only come in white.

These orchids are frequently seen in the tropics and most enjoy the steamy lowlands of Southeast Asia. They occur as epiphytes and the monopodial plants consist of only a few leathery leaves. These are often deep green, but some species have tessellated and mottled foliage.

The Philippines is the centre of distribution for the

Below:
Phalaenopsis stuartiana

Opposite

Top left:
Phalaenopsis schilleriana

Top right:
Phalaenopsis bellina

Centre left:
Phalaenopsis violacea

Bottom:
Phalaenopsis sumatrana

genus *Phalaenopsis*. Two of the most important species for producing pink flowered hybrids are the large flowered ***Phalaenopsis schilleriana*** and the smaller flowered ***Phalaenopsis equestris***.

Many of the spotted hybrids have been derived from ***Phalaenopsis stuartiana***, a species that also has attractive mottled leaves. ***Phalaenopsis lindenii*** is one of the miniature species, which has played a significant part in the breeding of the "candy stripe" line of *Phalaenopsis*.

Phalaenopsis javanica has only ever been recorded from west Java. It is now feared that this species is extinct in the wild, due to over-collection.

Another rare species is the delicate ***Phalaenopsis fimbriata***, being only known from a few sites in Indonesia.

The boldly marked ***Phalaenopsis sumatrana***, despite its specific name, is quite widespread throughout Southeast Asia. ***Phalaenopsis bellina*** was formerly well known as the Borneo form of ***Phalaenopsis violacea***.

Renanthera and Rhynchostylis

Opposite

Top left:
Renanthera monachica

Top right:
Rhynchostylis gigantea

Bottom left:
Rhynchostylis gigantea

Bottom right
Rhynchostylis retusa

Renanthera

Some 15 species of these tall and climbing plants are distributed from tropical Asia, to New Guinea and the Solomon Islands. This is a distinctive group. Its plants are characterised by their large, upright monopodial growth habit, often with branched flower spikes of numerous yellow to orange and red blooms. Many intergeneric hybrids have been made with *Renanthera*. Many of these are popular as cut flowers in Thailand, Singapore and the Philippines.

Renanthera monachica is a lowland species from the Philippines. It is a shorter and more compact species than most, yet still produces an impressive branched spike of flowers almost continually throughout the year. The striking flowers are bright yellow to orange-yellow, with dark red spots.

Rhynchostylis

These lowland vandaceous plants are known as "Foxtail Orchids" because of their densely flowered inflorescences. *Rhynchostylis* is a small genus of monopodial plants with only four or five members, closely related to *Aerides*.

Rhynchostylis gigantea from Thailand and Indochina is arguably the most popular species. It comes in a range of colours, from white with pink spotting and blotching, through various shades of purple to red. There are also bi-coloured forms and pure white strains. These have been further developed and multiplied by nurserymen. These orchids are often grown in teak baskets, with the thick, fleshy roots attaching to the timber or allowed to ramble into thin air. ***Rhynchostylis retusa*** is more widespread than the previous species, being found throughout Southeast Asia. It has long pendulous inflorescences of up to 60 white flowers with pink markings. This also comes in a white-flowered form, which has only recently become available in cultivation. These colour forms are stable and have been multiplied from seed.

Trichoglottis

Below:
Trichoglottis australiensis

Opposite

Top left:
Trichoglottis brachiata

Top right:
Trichoglottis geminata

Bottom left:
Trichoglottis philippinense

Bottom right:
Trichoglottis seidenfadenii

The vandaceous genus *Trichoglottis* comprises about 60 species whose centre of distribution is Indonesia and the Philippines. There is also a very rare species from north-eastern Australia—***Trichoglottis australiensis***. These are mostly upright monopodial plants which can grow quite tall with age. The leaves are very leathery and stiff, helping the plant tolerate full sun in the tropics. These orchids produce roots along nodes on the stem which help them attach to their host.

Some of the more horticulturally attractive species frequently seen in cultivation are the Philippine duo, the stunning velvet-purple ***Trichoglottis brachiata*** and the closely related, but distinct, ***Trichoglottis philippinensis***. Another Philippine species, ***Trichoglottis geminata***, has the added bonus of a delightful fragrance. This species was formerly known as *Trichoglottis wenzelii*. A seldom-seen species, having been only recently described, is ***Trichoglottis seidenfadenii***, from Thailand and Vietnam.

Above:
Vanda tricolor
Below:
Vanda javierae
Opposite
Top left:
Vanda coerulea
Top right:
Vanda coerulea
Centre left:
Vanda liouvillei
Bottom:
Vanda bensonii

Vanda

This is one of the most important genera of plants for cut flower production in Thailand. A large export industry has developed using a handful of species in an extensive hybridising program. Vandas are often seen grown in large outdoor beds in lowland botanical gardens, such as those in Thailand, Singapore and the Philippines.

One of the best known species is the spectacular blue ***Vanda coerulea*** from the higher mountains of India, through to northern Thailand. It is known for its cold tolerance, and some nurseries stock seedlings of line-bred plants to cater for orchid growers in temperate climates. Blue is a rare colour in flowers, and one that is even scarcer in orchids. *Vanda coerulea* has been used in numerous hybrids, both within *Vanda* and with related genera. Fortunately this unique blue colouration is transmitted to a high percentage of its progeny. Some of the more choice clones have deep blue tessellations throughout the flower.

Below:
Vanda roeblingiana

Opposite

Top left:
Vanda testacea

Top right:
Vanda roeblingiana

Centre left:
Vanda lilacina

Bottom:
Vanda luzonica

Vanda bensonii is not a frequently seen species. It is from Myanmar and Thailand and has an upright inflorescence of up to 14 tessellated brown flowers with a lilac labellum.

Vanda tricolor is a distinctive common species, found on rocks or trees on the fringes of lowland forest in Java, Indonesia. It has perfumed flowers.

One of the smaller members of this genus is ***Vanda testacea***. This light yellow and blue flowered species was formerly known as *Vanda parviflora*. It has a wide distribution, being recorded from Sri Lanka and India through to Thailand.

From lowland areas of the Philippines comes ***Vanda luzonica***. It has pretty white flowers with varying amounts of pink to purple splashing and blotching. The Philippine ***Vanda roeblingiana*** is an unusual species with brown flowers in various shades, and a bizarre anchor-shaped labellum. This orchid enjoys cooler conditions than the majority of the fifty or so species.

The following two species from Thailand have never been common in cultivation. ***Vanda lilacina*** has pale lilac-coloured small blooms, with a darker labellum.

In some forms, the segments are almost white. ***Vanda liouvillei*** is often erroneously labelled as *Vanda brunnea*.

Below
Vanda testacea

It produces long infloresecenses of predominantly brown flowers. The bifurcated labellum is distinctive.

Another very rare Philippine endemic is the closely related ***Vanda javierae*** with its snow white flowers. This species was only discovered and named in the 1980s, from a small patch of montane forest in central Luzon. It immediately caused a sensation among orchid enthusiasts. Due to its rarity, mature plants still fetch several hundred dollars each. Fortunately, these plants will become more affordable in the near future, as it has been successfully propagated from seed. It is, without question, one of the world's most impressive orchid species.

Index

Index